MW00452615

Plants

Growing and Using Energy

by Kate Boehm Jerome

Table of Contents

DEVELOP LANGUAGE

Plants grow in many different places. Plants can be different sizes and shapes.

Discuss the plants you see on these pages with questions like these:

Which plant grows fruit that people eat?

The _____ grows _____.

Where does the water lily grow?

The water lily _____.

Where is the dandelion growing?

_______________________________.

Tell how these plants are alike. Tell how they are different.

banana tree
grape hyacinths
bananas

Plants can have many different sizes and shapes, but every plant is either a **nonvascular plant** or a **vascular plant**.

Nonvascular plants, such as mosses, cannot grow very tall. However, some vascular plants, such as redwood trees, can grow as tall as a thirty-story building.

nonvascular plant – a plant that does not have special tissues to move water, minerals, and food through the plant

vascular plant – a plant with special tissues to move water, minerals, and food through the plant

Vascular plants grow taller than nonvascular plants because they have special tissues called **xylem** and **phloem.**

Xylem and phloem form tube-like pathways in vascular plants. These pathways run through the roots, stems, and leaves. Water, **minerals**, and food move through these pathways to all parts of the plant.

Xylem moves water and minerals from the soil up through the plant. Phloem moves food, usually made in the leaves, through the plant.

redwood tree

xylem – the tissue in vascular plants that moves water and minerals through the plant

phloem – the tissue in vascular plants that moves food through the plant

minerals – natural substances that plants need to grow

Plant Part	Job
Roots	• take in water and minerals • help hold the plant in place
Stems	• support the plant • carry water, minerals, and food between leaves and roots
Leaves	• make food for the plant

KEY IDEA Vascular plants have tissues that allow them to grow taller than nonvascular plants.

Seed-Making Plants

A healthy plant will grow bigger and make new plants. For many plants, the **seed** is the part that grows into a new plant. Most of the vascular plants you see every day are seed-making plants. These plants are called **gymnosperms** and **angiosperms**.

Have you ever seen a pinecone? A pinecone carries the seeds of the most common kinds of gymnosperm, the **conifers.** A spruce tree is a conifer. So is a pine tree.

seed – a plant part that can grow into a new plant

gymnosperms – seed plants that do not produce flowers

angiosperms – seed plants that produce flowers and fruits

conifers – gymnosperms that produce seeds on cones

▶ **Spruce trees are conifers.**

pinecones

SHARE IDEAS **Explain** why a pinecone is important to a conifer.

Angiosperms are vascular plants that need flowers to make seeds. After an angiosperm flower blooms, most parts of the flower die and fall away. However, the flower part that holds the seed changes into a **fruit.**

The fruit protects the angiosperm's seeds while they grow. When the fruit is ripe, the seeds are ready to grow into new plants.

fruit – a covering that protects angiosperm seeds

KEY IDEA Gymnosperms and angiosperms are plants that produce seeds.

A cherry tree is an angiosperm that produces flowers and fruits.

Parts that Protect

Plants cannot run away from animals that might eat them. Their roots hold them in one place. However, plants do have **defenses,** or ways to protect themselves.

Some plants have special parts, such as thorns or spines, which keep them from being eaten. If a plant can keep an animal from eating it, the plant has a chance to grow bigger and make seeds. Then more plants can grow.

defenses – ways to protect oneself

◀ **This ocotillo plant has big thorns to protect it.**

KEY IDEA Plants have defenses that help them survive.

YOUR TURN

CLASSIFY

Look at the pictures. They show parts of plants. Classify by telling whether each part belongs to an angiosperm or a gymnosperm.

Can you think of other plant parts that both angiosperms and gymnosperms might have?

plant part	gymnosperm	angiosperm
flowers		✓
fruit		
pinecone		

MAKE CONNECTIONS

Think about a place you have been to that has lots of plants. Draw a picture of that place. Label any plants you know.

USE THE LANGUAGE OF SCIENCE

What jobs do roots do for a plant?

Roots hold a plant in place. They also take in water and minerals from the soil.

CHAPTER 2

Plants Make Their Own Food

Plants need food to live and grow. They do not eat food like animals do. Instead, plants use energy from sunlight, or light energy, to make their own food. This process is called **photosynthesis.**

Photosynthesis happens only in cells that have **chlorophyll.** Chlorophyll gives plants their green color. It is the substance that allows plants to use light energy from the sun.

photosynthesis – the process in which plants use energy from sunlight to make food

chlorophyll – the substance that gives plants their green color and allows plants to use energy from sunlight to make food

In most plants, photosynthesis takes place in the leaves. But sometimes this process occurs in other places. For example, photosynthesis takes place in the stem of a cactus.

No matter where it happens, photosynthesis cannot begin without light. That's why most plants will die if they are kept in a dark room. Without light, plants cannot make the food they need to grow and survive.

KEY IDEA Photosynthesis is the process in which plants use energy from sunlight to make food.

The Process of Photosynthesis

Plants need more than light for photosynthesis to take place. They also need water. The roots of a plant take in water from the soil. The water then travels through xylem tissue to the stem and leaves.

Plants also need **carbon dioxide**. Carbon dioxide is a gas in the air. Living things, such as animals and people, **release** carbon dioxide when they breathe. Plants **absorb**, or take in, this carbon dioxide and use it during photosynthesis.

carbon dioxide – a gas used during photosynthesis

release – give off or let go

absorb – take in

▶ **Roots take in water from the soil. Leaves absorb carbon dioxide from the air. Both water and carbon dioxide are used during photosynthesis.**

There are two main steps in photosynthesis. During the first step, light hits chlorophyll in a plant. The chlorophyll uses this light energy to break down water in the plant. In this process, **oxygen** and **hydrogen** are produced.

Most of the oxygen is released into the air. But the hydrogen is used in the second part of photosynthesis. Hydrogen combines with carbon dioxide to make sugar. This sugar is the plant's food.

oxygen – a gas that is produced during photosynthesis and released into the air

hydrogen – a gas that is produced during photosynthesis and is used in the second part of photosynthesis

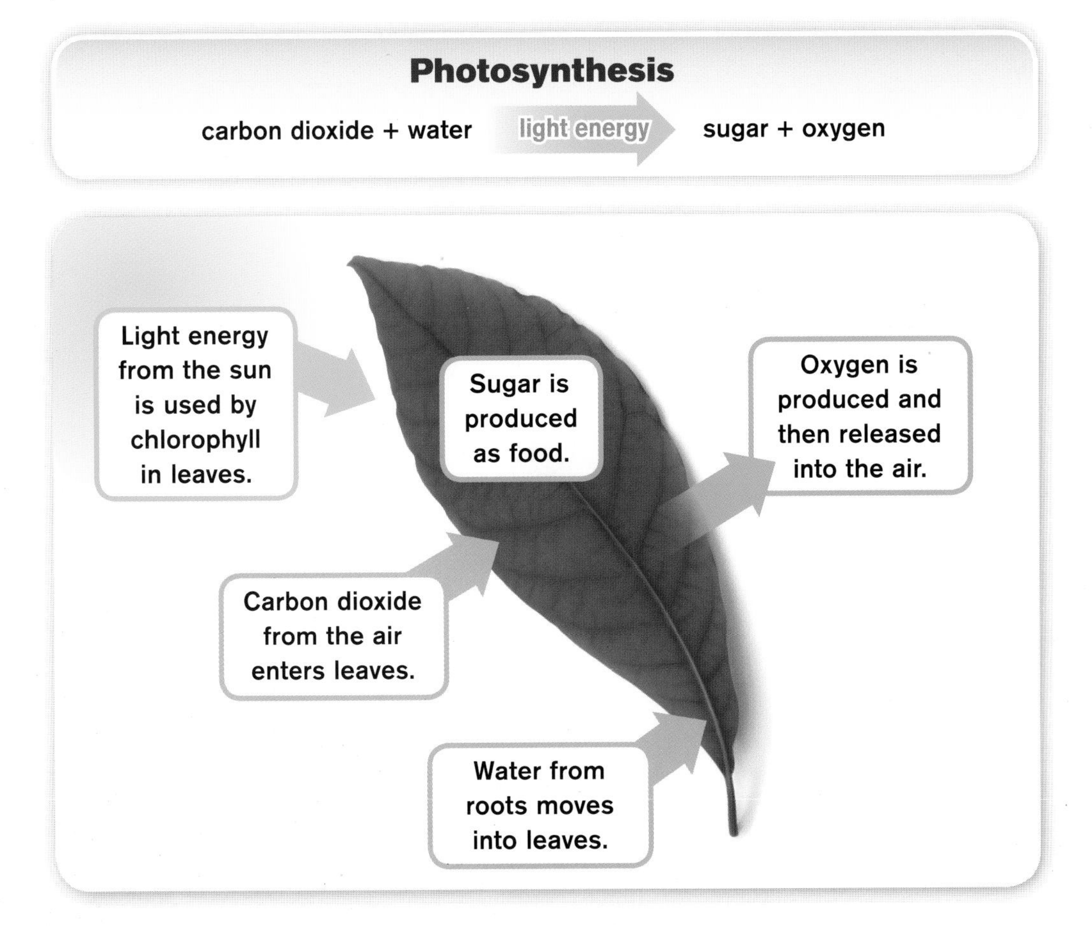

Plants Support Life

The process of photosynthesis helps explain why plants are so important. Through photosynthesis, plants produce their own food. This makes plants **producers**. As producers, plants provide food for animals and people to eat.

During photosynthesis, plants also release oxygen into the air. Most living things need this oxygen to survive.

Without plants, we would not have food to eat or air to breathe. They make it possible for us to live on Earth.

KEY IDEA During photosynthesis, light energy is used to change water and carbon dioxide into sugar and oxygen.

producers – living things that use light energy from the sun to make their own food

Plants release oxygen that is used by animals. Animals release carbon dioxide that is used by plants.

YOUR TURN

OBSERVE

Take a walk around your school. Make a list of places where plants are growing. Also look for places where plants are not growing. Think about what plants need to live. Then try to answer these questions.

1. Why do you think plants are growing in certain places?
2. Why do you think plants are not growing in other places?

MAKE CONNECTIONS

Many of the trees on Earth grow in rainforests. Can you give some reasons why these rainforests should be protected?

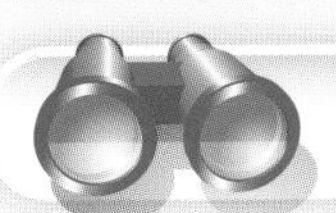

STRATEGY FOCUS

Make Connections

Look back at this chapter. What connections can you make to plants? Make a chart like this one. Write down your connections.

The text says . . . or The picture shows . . .	This reminds me of . . .	It helps me understand that . . .
(p. 14) a bear eating plants	a nature show I saw on TV	animals need plants to survive

CHAPTER 3

Plants Use Food for Energy

You know that plants make food during the process of photosynthesis. Plants use this food for energy. The process in which plants use the energy stored in food is called **respiration**.

Not all of the oxygen from photosynthesis is released into the air by a plant. Some of this oxygen is used for respiration. During respiration, stored sugar and oxygen are changed. Energy, carbon dioxide, and water are produced.

respiration – the energy-releasing process in plants

You could think of respiration as the opposite of photosynthesis. Photosynthesis uses energy, carbon dioxide, and water to produce sugar and oxygen. Respiration uses sugar and oxygen to release energy and produce carbon dioxide and water.

There are other differences between photosynthesis and respiration, too. Photosynthesis can occur only when light is present. But respiration does not need light, so it can occur in darkness or light.

Another difference is that respiration can occur in any plant cell. But photosynthesis can occur only in cells that have chlorophyll.

Photosynthesis	Respiration
• occurs only in cells with chlorophyll	• occurs in all cells
• energy is stored in sugar	• energy is released from sugar
• carbon dioxide is used	• carbon dioxide is produced
• oxygen is produced	• oxygen is used

You know that plants produce oxygen in the process of photosynthesis. But you also know that plants use oxygen in the process of respiration. How does this affect the amount of oxygen in the air?

Luckily, plants produce more oxygen during photosynthesis than they use during respiration. This means there is enough oxygen in the air for other living things to breathe!

By The Way...

Some plants lose their leaves in the fall. In winter, the plants survive by using the food they stored during the summer.

Plants produce the oxygen that people and other animals breathe.

KEY IDEAS Plants get energy from stored food in a process called respiration. Respiration can occur during the day or night.

YOUR TURN

INFER

Look at the pictures of the plants. Then choose the best word to infer what process can be occurring in each plant.

1. Photosynthesis can/cannot be occurring
2. Respiration can/cannot be occurring

1. Photosynthesis can/cannot be occurring
2. Respiration can/cannot be occurring

MAKE CONNECTIONS

Cells in animals also use respiration to get energy from food. How do you think animals get the oxygen they need for respiration?

EXPAND VOCABULARY

The prefix *photo* means "light." Find out about these words.

photograph **photocopy** **photometer**

What do these words have to do with light? Explain with words, diagrams, or drawings.

Botanical Gardens

In a botanical garden, people can enjoy many different kinds of plants and learn about them. Many people are needed to take care of all the beautiful plants.

Some workers keep plants healthy. Some people learn how to do this by training on the job. Other people, such as plant researchers, learn about plants in college.

Some workers teach visitors about plants and explain why they are important. This is a good job for people who enjoy talking to others and answering questions about plants.

- Would you like to work at a botanical garden?
- Why or why not?

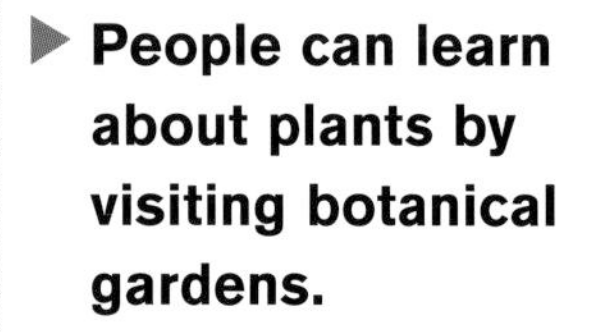

▶ People can learn about plants by visiting botanical gardens.

Verbs that Persuade

Some verbs are used to make requests or give commands. They can also be used to persuade. These verbs are called imperative verbs.

EXAMPLE

Look at that wonderful tree!
Give me the camera.
Let me take your picture. **Smile**!

Page through the book. Give commands to a friend, such as "Point to the cherry tree" or "Read the caption on page 8." Use imperative verbs.

Write a Speech

Someone in your town wants to cut down a tree on your block. You love that tree! Persuade your friends to join you in your fight to save the tree.

- Tell why the tree is important.
- Mention people and animals that are affected.
- Use imperative verbs.

Words You Can Use	
listen	save
don't	stop
sign	tell

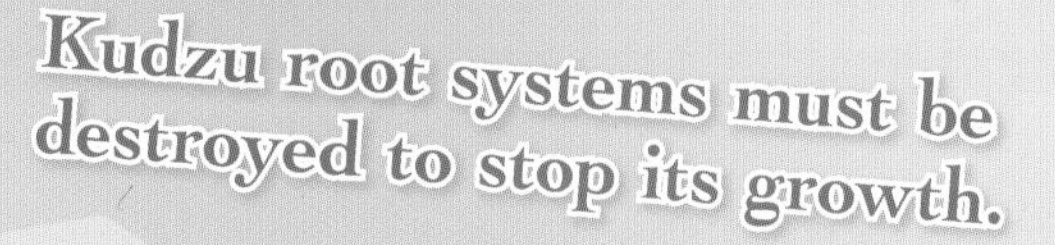

Sometimes plants are brought into an area and start to grow out of control. Kudzu is a vine that was first brought to the United States in 1876. It now grows out of control in some areas in the Southeast.

Read the poster and answer these questions.

- Why is kudzu a problem?

- How can the growth of kudzu be stopped?

Key Words

angiosperm (angiosperms) a seed plant that produces flowers and fruits
Any plant that produces flowers is an **angiosperm**.

carbon dioxide a gas used during photosynthesis
Plants take in **carbon dioxide** to make food.

chlorophyll the substance that gives plants their green color and allows plants to use energy from sunlight to make food
Plants usually have **chlorophyll** in their leaves.

fruit (fruits) a covering that protects angiosperm seeds
An orange is a **fruit** that contains many seeds.

gymnosperm (gymnosperms) a seed plant that does not produce flowers
A pine tree is a **gymnosperm** that produces seeds on cones.

nonvascular plant (nonvascular plants) a plant that does not have special tissues that move water, minerals, and food through the plant
Moss is a **nonvascular plant**.

oxygen a gas that plants release into the air during photosynthesis
Plants release **oxygen** during photosynthesis.

phloem the tissue in vascular plants that moves food through the plant
Phloem runs through the roots, stems, and leaves of vascular plants.

photosynthesis the process in which plants use energy from sunlight to make food
Plants make food during the process of **photosynthesis**.

respiration the energy-releasing process in plants
Plants release the energy in stored food during **respiration**.

seed (seeds) a plant part that can grow into a new plant
Angiosperm **seeds** are covered by fruit.

vascular plant (vascular plants) a plant with special tissues that move water, minerals, and food through the plant
A **vascular plant** has xylem and phloem tissues.

xylem the tissue in vascular plants that moves water and minerals through the plant
Xylem carries water and minerals through a vascular plant.

Index

MILLMARK EDUCATION CORPORATION
Ericka Markman, President and CEO; Karen Peratt, VP, Editorial Director; Rachel L. Moir, Director, Operations and Production; Mary Ann Mortellaro, Science Editor; Amy Sarver, Series Editor; Betsy Carpenter, Editor; Guadalupe Lopez, Writer; Kris Hanneman and Pictures Unlimited, Photo Research

PROGRAM AUTHORS
Mary Hawley; Program Author, Instructional Design
Kate Boehm Jerome; Program Author, Science

BOOK DESIGN Steve Curtis Design

CONTENT REVIEWER
Nikki L. Hanegan, PhD, Brigham Young University, Provo, UT

PROGRAM ADVISORS
Scott K. Baker, EdD, Pacific Institutes for Research, Eugene, OR
Carla C. Johnson, EdD, University of Toledo, Toledo, OH
Donna Ogle, EdD, National-Louis University, Chicago, IL
Betty Ansin Smallwood, PhD, Center for Applied Linguistics, Washington, DC
Gail Thompson, PhD, Claremont Graduate University, Claremont, CA
Emma Violand-Sánchez, EdD, Arlington Public Schools, Arlington, VA (retired)

PHOTO CREDITS Cover © Darren Matthews/Alamy; 1 © Jason Keith Heydorn/Shutterstock; 2a © Nick Kirk/Alamy; 2b and 18b © blickwinkel/Alamy; 2-3 © vera bogaerts/Shutterstock; 3a © Tropicalstock.net/Alamy; 3b © James Clarke Images/Alamy; 4 © Bob Stefko/The Image Bank/Getty Images; 5 © Edward Parker/Alamy; 6a © Robert McGouey/Alamy; 6b © Arco Images/Alamy; 7a and 7b © AGStockUSA, Inc./Alamy; 8 © Lighthouse Imaging/Alamy; 9a Deborah Aronds; 9b © Billy Lobo H./Shutterstock; 9c © Elnur/Shutterstock; 9d and 9e Lloyd Wolf for Millmark Education; 10-11 © Danita Delimont/Alamy; 12 © Design Pics Inc./Alamy; 13 illustration by Joel and Sharon Harris; 14 © Keren Su/Getty Images; 15a © Lynne Furrer/Shutterstock; 15b © Kathy Burns-Millyard/Shutterstock; 16-17 © Carlos Davila/Alamy; 18a © Losevsky Pavel/Shutterstock; 19a © Image Source Black/Getty Images; 19b © Telnova Olya/Shutterstock; 20 © Ray Kachatorian/Getty Images; 21 © Jin Yong/Shutterstock; 22b © Danny E. Hooks/Shutterstock; 22a © fenix rising/Alamy; 24 © Jakez/Shutterstock

Copyright © 2008 Millmark Education Corporation

All rights reserved. Reproduction of the whole or any part of the contents without written permission from the publisher is prohibited. Millmark Education and ConceptLinks are registered trademarks of Millmark Education Corporation.

Published by Millmark Education Corporation
7272 Wisconsin Avenue, Suite 300
Bethesda, MD 20814

ISBN-13: 978-1-4334-0044-5
ISBN-10: 1-4334-0044-8

Printed in the USA

10 9 8 7 6 5 4 3 2 1